Butter-Making Day

Mari Purdy
Illustrated by Tranda Strnad

Prairie Flower Publishing

Butter-Making Day
Copyright © 2025 by Mari Purdy

ISBN: Hard Cover (11" × 8.5") – 979-8-9934728-0-5
ISBN: Soft Cover (11" × 8.5") – 979-8-9934728-1-2
ISBN: Soft Cover (8.25" × 6") – 979-8-9934728-2-9

Prairie Flower Publishing

Prairie Flower Publishing
1239 N. Krug St.
Russell, Ks. 67665
maripurdy@icloud.com

This story is about my friend, Susan Louise Marlatt Krug, and her life on the family farm in Atchison County, Kansas. Thank you, Susie, for sharing your story.

My Friend, Susan

Susan loved her life on the farm. She learned many things and had lots of fun.

In addition to working the land, her family raised cattle and chickens. Life wasn't easy, but they all worked together to prosper the best they could.

This story is about Susan's job of milking the cow and making butter. Old Bossy was her milk cow, Sugar was her horse, and Toby was the family dog.

Susan and her sister, Priscilla Ann, still make extended visits to their family farm. Those visits help Susan to remember, and those memories become stories—stories about her family heritage.

Her dream is for her two children and five grandchildren to hear the tales of her past. This is just one of Susan's many stories.

This book belongs to:

my name is Susan Louise Marlatt. I attend country school in Atchison County, Kansas. My teacher's name is Mrs. Duffy. We have twenty-five students in our class. We have all grades, first through eighth, in our one-room school. Mrs. Duffy teaches all of us, and we teach our friends. It is fun learning at Terry School.

When school is out, my sister, Priscilla Ann, and I run to get Sugar from the school horse pen. Sugar is my horse. She neighs as she comes running to see us. I take her bridle off the fence and put it on her. Priscilla Ann and I ride her down the winding road to our farm.

The closer we get to home, the faster Sugar goes. She is in a hurry to get home to eat the sweet green grass, but we don't care because…today is butter-making day! Priscilla Ann and I are excited to get home and get our chores done. Butter-making day is always a special day on our farm. On that day, Mother surprises us with something to eat with our freshly made butter. She will make pancakes, or homemade bread or buttermilk biscuits.

I start calling for Old Bossy as soon as I see her tail swishing in the tall alfalfa. "Sooooo, Bossy, Sooooo, Bossy!" She knows when I start calling her that it is time to come to the barn for her evening milking and her fresh treat of hay.

Sugar, Priscilla Ann, and I run into the barnyard just as Old Bossy lumbers in from the pasture. We wave to Daddy as we gallop by. I hop off Sugar and open the door of the barn to let Old Bossy in. While I give her some hay to make her happy, Priscilla Ann takes off Sugar's bridle and turns her out in the pasture. Sugar runs and bucks as she races toward the green grass. That's her way of saying she's happy to be home. Priscilla Ann goes to gather the eggs, and I begin my job of milking Old Bossy.

milking the cow is a big responsibility, but my daddy reassures me saying, "Susan Louise, now that you are in the fourth grade, I trust you to do the milking." I proudly get my bucket from the milkhouse and find my stool to sit on. My daddy made it for me. It is two boards nailed together to form a "T." It is very hard to sit on. However, once my knees are around the bucket and my head rests on Old Bossy's warm belly, the job becomes easy.

My Daddy had taught me how to milk a cow. The more I practiced, the better I became. My fingers and hands grew very strong.

The old barn cat hangs out while I'm milking and rubs against my legs. Her purring is in harmony with the rhythmic swishing sound the milk makes as it drops into the bucket.

I give the cat a taste of fresh milk. It doesn't take too much practice to spray milk that far. I can even write my name in cursive on the barn wall with milk. "Meow," the old cat replies in thanks as she licks that nice warm milk off her fur.

Toby, our farm dog, sits by the stall and whines in anticipation. He knows if he hangs around, he might get a drink of milk too.

Sometimes, I even give myself a squirt of warm milk. Yummy! There is nothing better than Old Bossy's fresh, warm milk.

As soon as my bucket is full, I set it in the milkhouse and turn Old Bossy back out to the pasture. She heads to the creek to get a drink. I carry the bucket of milk to the house for mother to pour into a gallon jar.

I see my mother's smiling face as I hand her the milk. Priscilla Ann gives her eight eggs.

"What!" Mother exclaims. "You have your chores done already?"

We spot the butter churn on the counter, so we know Mother remembers today is butter-making day.

Mother puts the milk in the refrigerator to cool. Then, she takes out a cold gallon of milk from yesterday's milking. The thick cream has separated from the milk and lays in a gooey layer on top.

"Wash your hands," Mother says. She gives me a ladle to dip the cream off the milk while Priscilla Ann gets the churn ready. I scoop the smooth, rich, creamy liquid out of the jar and drop it into the butter churn.

We take the butter churn to the back porch where Priscilla Ann and I take turns cranking the handle. When we first begin churning the cream, the handle is very easy to turn. That's why Priscilla Ann always wants to go first. Soon, the cream becomes foamy. We keep turning the handle. When the cream begins to thicken, I take over. It is fun to see the lumps of butter beginning to form. The lumps get bigger and bigger until they start sticking together. The butter begins to turn a yellowish color. The white liquid that separates from the butter is called buttermilk.

When the butter has formed into a solid clump, Mother takes the churn to the kitchen. She removes the lid and pours everything into a strainer. She pushes all the buttermilk out of the butter with a wooden spoon. Then, she pours the buttermilk into a pint jar. It will be used for drinking or baking.

mother then puts the butter into a big bowl to be washed. We use ice water and a wooden spoon to work the butter until all the buttermilk is washed out. Soon, it is nice and creamy. Mother adds a bit of salt. We pat it dry and place the butter in the butter crock.

After cleaning up our mess and washing the butter churn, Mother has Priscilla Ann and me wait for her at the picnic table on the back porch. She surprises us with fresh, warm, buttermilk biscuits slathered all over with butter and a big glass of Old Bossy's cold milk. It is our favorite after-school snack.

This is the best part of butter-making day on the farm! Yummy!

BUTTER

Wish
List
Farm

Thank you, Old Bossy, for sharing your cream with us!

Milk comes from cows,

Cream comes from milk,

And butter comes from cream!

Yum!

How to Make Butter

Things you will need:

2 pint jars

½ cup of heavy cream (room temperature)

Strainer

Large bowl

Wooden spoon

Small pitcher of ice water

Salt

Paper towel

Butter dish

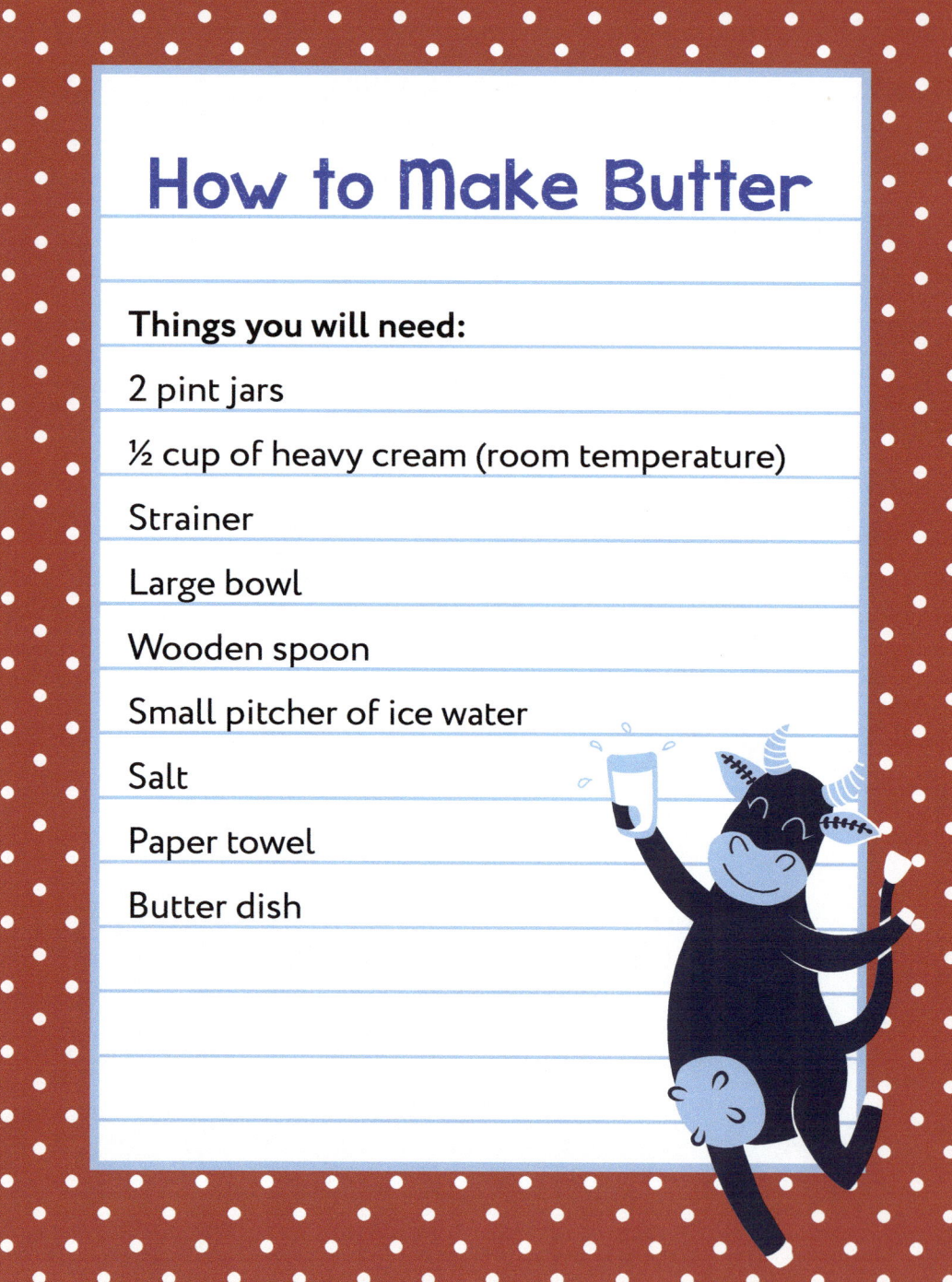

1. Wash the pint jars and add ½ cup of heavy cream to one.

2. Screw on the lid and make sure it is tight.

3. Shake the jar until it forms a solid lump of butter. The buttermilk will separate from the butter. (This takes 10-15 minutes.)

4. When the butter is lumped together, dump the butter and buttermilk into the strainer. After the butter-milk runs through the strainer, pour it into the second jar. Put the butter in a large bowl.

5. Rinse the butter with ice-cold water. Using a wooden spoon, push the butter-milk out of the butter.

6. Add a dash of salt to the butter and work it in.

7. Pat the butter dry with a paper towel.

8. Put butter in a serving bowl or butter dish.

9. Eat it!

www.ingramcontent.com/pod-product-compliance
Lightning Source LLC
Chambersburg PA
CBHW040818120626

46551CB00004B/598